The Future Equation:
Balancing Humanity and Technology

A Collection of Essays

by

Mekiki

Mekiki Magazine

First published in English in 2023 By Mekiki Magazine

Copyright for the first English edition © 2023 Mekiki Magazine

Site: www.mekikimagazine.com

ISBN: 978-1-954145-61-0

Title: The Future Equation: Balancing Humanity and Technology

Author: Mekiki
Editors: Mekiki and A. Lee
Book Cover By: Álvaro Oliveira For Mekiki Magazine
Graphic Design: Álvaro Oliveira For Mekiki Magazine

Series Foreword

I n the present digital age, an era characterized by immediate access to a boundless array of information, we find ourselves in an ocean of opinions. While this wealth of data may satiate our curiosity on a superficial level, it often fails to provide the depth of understanding we crave—the kind of fundamental understanding that gives birth to a conscientious and comprehensive perception of the world. This is precisely where Mekiki Magazine comes into play. We stand as a beacon in the information storm, offering not just information but knowledge, wisdom, and insight.

At Mekiki Magazine, we believe in intertwining cutting-edge technology with the wisdom of industry experts to curate highly tailored content. With the integration of our unique proprietary content, we are revolutionizing how knowledge is disseminated, making it much more accessible globally. Our innovative approach harnesses technology to deliver insights that cater to global preferences and needs. We are committed to empowering our writers with advanced intelligence capabilities, both generative and predictive, to enhance their understanding of the interests of our readers. This, in turn, allows them to deliver content that truly resonates with their audience, providing unparalleled depth and value. At Mekiki Magazine, we are not just about informing; we are about transforming understanding, one reader at a time.

Our mission is to break down the walls between specialists and the broader audience, making all topics accessible and engaging to all. Each of our publications, whether a magazine, book, or online article, acts as a unique portal into realms of knowledge. We are not simply content creators but translators of intricate subjects, transforming abstract ideas into comprehensible, engaging narratives. In essence, we aim to fill the gap between information and understanding. At Mekiki Magazine, our readers do not just learn—they comprehend. They do not just read—they engage. And in doing so, they gain a richer, more nuanced view of the world around them.

Join us on a transformative journey of discovery with Mekiki Magazine, where cutting-edge technology merges with timeless wisdom to illuminate new horizons of knowledge and understanding.

Contents

The Current State of Humanity 1

The Impact of Emerging Technologies and Global Challenges 3

Purpose of the Book 6

The Rise of Technology: *Artificial Intelligence* 8

Essay 1 *The Evolution and Future of AI* 10

Essay 2 *Ethical Challenges Posed by AI* 13

Essay 3 *AI and Job Displacement* 16

Essay 4 *AI in Health Care* 18

Essay 5 *VR Beyond Games* 20

Essay 6 *VR and Social Interaction* 22

Essay 7 *VR and Mental Health:* Potential and Pitfalls 24

Essay 8 *VR in Future Workplaces:* Prospects and Challenges 26

Essay 9 *Privacy in the Age of Big Data:* An Uncharted Frontier 30

Essay 10 *Surveillance Technologies and Society:* A Careful Balancing Act 32

Essay 11 *Balancing National Security and Privacy:* Striking the Delicate Equilibrium 35

Essay 12 *The Role of AI in Cybersecurity:* An Emergent Guardian 38

Essay 13 *Social Media's Role in Modern Democracy* 41

Essay 14 *Emerging Models of Democratic Governance* 44

Essay 15 *Human Rights in the Digital Age* 47

Essay 16 *Human Impact on Climate* 52

Essay 17 *Effects on Biodiversity and Global Health* 55

Essay 18 *Socioeconomic Impact and Sustainable Future* 58

Essay 19 *The Era of Genetic Engineering* 61

Essay 20 *Ethical Considerations of Gene Editing* 64

Essay 21 *Biotech's Role in Food Security and Agriculture* 67

Essay 22 *The Silent Epidemic of Mental Health* 72

Essay 23 *Social Media and Mental Health* 75

Essay 24 *Mental Health in the Workplace* 78

Essay 25 *The Future of Mental Health Treatment* 81

Essay 26 *The Widening Gap Between Rich and Poor* 84

Essay 27 *Automation's Impact on Income Inequality* 87

Essay 28 *Income Inequality and Health Disparities* 90

Essay 29 *Policy Solutions to Combat Income Inequality* 93

Essay 30 *The New Space Age and Mars Colonization* 98

Essay 31 *Ethical Considerations of Space Exploration* 100

Essay 32 *The Implications of Discovering Extraterrestrial Life*........................ 103

Reflections on Current Themes Shaping Humanity................................. 106

    The interconnectedness of themes explored.................................108

    A call to action for addressing these challenges............................. 110

Concluding Remarks 112

Glossary 115

The Current State of Humanity

◄♦━━━━━━━━━━━━━━━━━━━━♦►

In contemplating the state of our world, it is essential to underscore the paradox that defines our era. This paradox lies in the remarkable advancements we have achieved in science and technology and the deep societal fissures that are simultaneously widening. In this dissonance between progress and division, we find the crux of our exploration.

We live in an era of unprecedented innovation. There is a palpable sense of change in the air—a sort of metamorphosis that humanity has not experienced in previous epochs. Artificial intelligence (AI), once a concept relegated to the realm of science fiction, is becoming integral to our daily lives. We are harnessing the power of the atom, delving into the mysteries of the human genome, and peering into the far reaches of space with ever-increasing clarity.

Yet beneath this veneer of progress, we grapple with a series of complex, interconnected issues. Our global society contends with burgeoning inequality, struggling to reconcile the abundance of wealth with deep-rooted poverty. Climate change, a specter of our creation, looms ominously, challenging our collective survival. The importance of privacy and security concerns is becoming increasingly clear in a world where the digital web has created a hyperconnected society.

This is the juxtaposition that defines our time: a world teetering on the brink of unprecedented potential and catastrophic failure. It is a delicate balancing act, and how we choose to act now will shape the course of humanity for generations to come.

Our exploration is not academic. We are seeking an understanding that goes beyond the surface, diving into the heart of our challenges and possibilities. This demands a rigorous analysis and a search for the "why" that underpins the "what." We are in pursuit of a comprehensive, well-rounded picture that can inform meaningful action.

By peering through the lens of various disciplines—technology, politics, sociology, and more—we hope to glean insights that inform this understanding. The ultimate goal is not simply to paint a picture of the current state of humanity but to spark a discourse that can guide us toward a better future.

In this complex dance between progress and challenge, between innovation and consequence, we find the narrative of our time. Our actions and choices continue to influence and shape this ongoing story as it unfolds. We welcome you to join us on this journey of exploration as we explore further the intricate, sometimes troubling, but always fascinating realm of our shared humanity.

The Impact of Emerging Technologies and Global Challenges

W e find ourselves at an inflection point where the ceaseless march of technology meets the formidable landscape of global challenges. This crossroads forms the stage upon which humanity's future will be decided, and its players are emerging technologies and the increasingly complex problems of our world.

Consider the transformative power of artificial intelligence (AI), which is shaping industries, streamlining processes, and speeding up research. A marvel of human ingenuity, AI holds the potential to drive unprecedented efficiency and break the boundaries of knowledge. Yet it also prompts serious ethical questions regarding job displacement, privacy, and the very nature of human–machine interactions. Our challenge is not to determine whether AI can progress but to decide how it should progress, what moral principles should be upheld, and whose interests should be served.

Similarly, virtual reality (VR) extends its immersive hand beyond entertainment, reaching into education, therapy, and professional training. VR's capacity to bridge geographical divides and create empathetic experiences is powerful. However, concerns arise over psychological effects, social isolation, and the potential for misuse in spreading disinformation or propaganda.

In the realm of biotechnology, our ability to edit the genetic code places us on the precipice of eradicating inherited diseases and enhancing food security. Yet it also opens Pandora's box of ethical implications, from the danger of creating genetic inequality to the unforeseen consequences of biodiversity.

Parallel to the rise of technology, the specter of global challenges casts a long shadow. Climate change, a product of our industrial prowess, threatens ecosystems and human societies alike. The advancement of automation undermines social stability by increasing income inequality. Cybersecurity risks, spurred by our ever-increasing connectivity, pose threats to personal privacy and national security.

Thus, we find ourselves in a dance of progress and peril. Emerging technologies promise extraordinary benefits, but they also amplify existing challenges and bring about new dilemmas. Meanwhile, our global issues demand urgent action but also present opportunities for technological solutions.

What then becomes apparent is the need for a carefully considered approach. An approach that balances the promise of technology with a deep respect for the potential ramifications and views global challenges not as insurmountable obstacles but as catalysts for innovation and cooperation.

The journey to understand the current state of humanity depends on this delicate balancing act. As we unravel the complexities of each theme, we will be mindful of their interconnections and the ripple effects that connect technology, society, and our collective future. This understanding will serve not just as a map of where we are but also as a compass, guiding us toward where we should aspire to be.

Purpose
of the Book

◄◆━━━━━━━━━━━━━◆►

As we ponder the rapidly changing contours of our world, the aim of this volume is threefold. First, it seeks to examine the grand tapestry of our time to understand the threads of emerging technology and global difficulties that interweave to shape our collective existence. Second, it endeavors to ignite discussion, provoke thought, and encourage critical evaluation of our current trajectory. Third, it aspires to inspire, not dismay, by painting a picture of potential futures that we can strive toward.

The dialogue about our present state is abundant yet often compartmentalized. Discussions on technology exist in one sphere, and debates on global issues exist in another. This book dares to cross these boundaries, exploring the interconnections between these spheres and how they influence and shape one another.

In this exploration, we explore the possibilities and quandaries of artificial intelligence. We traverse the landscape of biotechnology, balancing its promise against the ethical considerations it raises. We examine the role of privacy and cybersecurity in our hyper-connected era and the fine line between security and the erosion of personal freedoms. And we scrutinize the multifaceted challenge of climate change, our responsibility toward it, and the radical transformations it may demand of us.

But this exploration is not intended to be a passive exercise. Rather, it is a call to engage in discourse and contemplate the implications of these technologies and challenges. The questions posed are not rhetorical, but real. Their answers are not predetermined but are awaiting our collective decision.

Moreover, the purpose of this book is not to predict doom or incite fear. Instead, it seeks to inspire and give an image of what could be if we navigate this complex terrain with wisdom, empathy, and foresight. It posits that while the problems are immense, so too are our capacities for innovation, adaptability, and resilience.

This book's goal is to contribute to the creation of a world that, faced with momentous change and challenges, opts for conscious evolution. A more in-depth understanding of the threads that form our present guides us in weaving a future that is inclusive, sustainable, and human-centric. This is a quest not only for understanding but also for imagination and aspiration. And it is a journey on which we invite you to join us.

The Rise of Technology:
Artificial Intelligence

◄◆━━━━━━━━━━━━━━━━━━━◆►

A rtificial intelligence, a concept once relegated to the realm of science fiction, is now an integral part of our daily lives. This chapter attempts to explicate the essence of this profound technological advance and its overarching impact on society.

We are at the dawn of a new age, one marked by rapid advancements in artificial intelligence. Whether it is self-driving cars navigating our city streets or virtual assistants managing our daily schedules, artificial intelligence has emerged as a transformative force in the modern world. Yet this rise of AI raises as many queries as it answers.

The advent of AI has unlocked unparalleled opportunities for businesses and individuals alike. It has redefined our productivity, enabling us to analyze vast amounts of data with stunning speed and accuracy, transforming industries from health care to finance, from education to transportation. Yet we must grapple with the stark reality that these machines, powered by complicated algorithms, can render many traditional jobs redundant.

AI has also invaded our personal lives, with algorithms predicting our preferences, managing our schedules, and even influencing our decisions. This evokes pertinent inquiries about the balance between convenience and privacy. Are we sacrificing our autonomy and privacy on the altar of AI-powered convenience? The answer, as is often the case with complex phenomena, is not clear-cut.

Our interaction with AI also triggers deeper philosophical questions about what it means to be human. As machines become increasingly capable of replicating and even surpassing human intelligence, we are compelled to reevaluate our understanding of cognition, consciousness, and the human condition.

In essence, artificial intelligence stands as a beacon of our technological prowess and a testament to our relentless pursuit of innovation. It also serves as a stark reminder of the dilemmas we face in this age of rapid technological advancement. As we strive to harness the immense potential of AI, we must also address the ethical, societal, and philosophical questions it raises, ensuring a future where technology serves humanity, not the other way around.

In the following essays, we will explore how AI is changing our world and consider how we can navigate this complex landscape.

Essay 1
The Evolution and Future of AI

At the heart of our advancing digital age pulses a force of extraordinary potential—artificial intelligence. Born of human imagination and refined by our ceaseless drive for growth, AI has developed from a theoretical concept into a transformative force.

AI's progress began with humble roots in the 1950s, a seed planted in the fertile ground of mathematical theory and computational science. However, technological limitations and the sheer complexity of simulating human intelligence tempered its initial promise. The ensuing years saw a rollercoaster of progress and setbacks, periods of optimism known as "AI springs" interspersed with "AI winters" of disillusionment.

The twenty-first century brought with it an explosion of computational power and data—the lifeblood of AI—propelling it out of its winter. As a result of advancements in neural networks and rising computational power, machine learning and subsequent broad learning came to the forefront. This new wave of AI, termed "narrow AI," excelled at specific tasks, from voice recognition to image analysis, transforming industries and daily life.

Looking ahead, the future progression of AI promises an era of "artificial general intelligence" (AGI), where AI can execute any cognitive task equivalent to that of a human. The vision is of systems that understand, learn, adapt, and implement knowledge across a broad range of tasks, a stark departure from our current "narrow AI." This leap would have serious implications, opening doors to unimaginable improvements but also posing substantial ethical, societal, and existential questions.

The probable benefits are immense. AGI could revolutionize health care by providing personalized treatment plans or aiding in complex surgeries. It could speed up research in fields ranging from climate science to quantum physics. It could enhance education by offering individualized learning experiences.

Yet the emergence of AGI also raises intense challenges. The risk of job displacement could heighten social inequality. Concerns about privacy and control could escalate. The potential for misuse in areas such as autonomous weapons is deeply troubling. There are also philosophical questions about what such advancements mean for our understanding of intelligence, consciousness, and our place in the universe.

In navigating the future of AI, we face a delicate balancing act. It is a course of pursuing the unique promise AI holds while also treading with serious caution. A journey of pushing the boundaries of what is technologically possible while also questioning what is ethically acceptable and socially desirable.

The growth and future of AI thus present us with a unique challenge. The path we choose in the face of this challenge will determine not just the future of AI but the future of humanity itself. It is a path that demands from us not just technological prowess but in-depth wisdom, ethical consideration, and collective decision-making. As we continue our exploration of AI, we do so with a keen awareness of its complexity and the profound responsibility it places on our shoulders.

Essay 2
Ethical Challenges Posed by AI

The digital revolution presents a challenge in terms of humane considerations. As a compass to navigate this terrain, our discussion turns to the ethical challenges inherent to the unfolding narrative of artificial intelligence.

The first quandary revolves around accountability. In an AI-driven world, determining responsibility for decisions made or actions taken by autonomous systems becomes murky. If a sovereign vehicle, taught to make split-second decisions, causes harm, who is to be held accountable? The programmer? The manufacturer? Or the AI system itself? This lack of clear responsibility can lead to moral and legal conundrums.

AI also grapples with issues of bias and fairness. Machine learning algorithms are only as impartial as the data they are trained on. If the training inputs carry biases, the AI will inherently adopt these prejudices, leading to discriminatory outcomes in areas like hiring, law enforcement, and loan approvals.

A third challenge involves privacy and surveillance. In an age where statistics are the new oil, AI's voracious appetite for information could lead to an erosion of personal privacy. The rise of facial recognition technologies and predictive policing, while potentially beneficial, could equally usher in a dystopian reality of unwarranted surveillance and intrusion.

The potential for AI and automation to disrupt the work market raises yet another challenge. As AI can perform increasingly complex tasks, many jobs may become redundant, exacerbating economic inequalities. Balancing the need for progress with the potential for widespread job displacement presents a complex conundrum.

Finally, the potential for AGI—AI that matches or surpasses human intellect—also poses significant dilemmas. Will AGI have rights? How do we ensure such systems align with human values and priorities? And how do we mitigate the risk of a "superintelligent" AI that might pose an existential risk to humanity?

Navigating these challenges will require not just technical acumen but deep reflection, informed public discourse, and judicious policymaking. Ensuring that AI serves the collective good rather than amplifying existing inequities or creating new ones is a critical task. We must ask not just what AI can do but what it should do.

In conclusion, the challenges posed by AI are as vast as they are intricate. They demand a thoughtful and concerted response from all stakeholders—from policymakers and technologists to educators and citizens. As we move deeper into the age of AI, our capacity to engage with these challenges will significantly shape the world that we are building for ourselves and future generations.

Essay 3
AI and Job Displacement

◄◆———————————◆►

In the ongoing chronicle of human technological innovation, few chapters instigate as much apprehension as the ability of artificial intelligence (AI) to displace jobs. The melding of sophisticated AI with intricate automation mechanisms could dramatically transform the landscape of production, labor, and employment.

Firstly, it is crucial to discern that the displacement by AI is not uniformly distributed across sectors or regions. Jobs composed of highly repetitive tasks or those that require vast data processing capabilities are more susceptible. In contrast, roles demanding creativity, emotional intelligence, or complex problem-solving are less likely to face an immediate shift.

However, the allure of efficiency and profit maximization might lead businesses to automate even semiskilled jobs. The World Economic Forum estimates that by 2025, machines and AI will perform more responsibilities than humans in the workplace. The transformative power of AI may amplify existing social and economic inequalities, with those in lower-skilled roles disproportionately affected.

We must ponder on the capabilities of AI-induced underemployment. Even if AI does not replace human activity entirely, it might decrease the demand for certain skills, leading to career degradation. Furthermore, if the proliferation of AI leads to an abundance of workers, employers could leverage this to suppress wages, worsening income inequality.

However, it is essential to recognize that AI could also generate new employment opportunities. As AI and automation become more integrated into our economies, there will be an increasing demand for professionals who can develop, maintain, and manage these technologies. AI might free up employees from mundane duties, providing possibilities for more meaningful work.

In conclusion, the interplay between AI and job displacement is intricate and multifaceted. It is not merely a question of how many jobs will be lost but also of how work, as a societal construct, will be transformed. As we continue to harness the power of AI, we must create strategies that mitigate potential work displacement, invest in reskilling, and build an economy that values all forms of labor. It is through this thoughtful approach that we can ensure a future in which humans and AI perform in concert rather than in competition.

Essay 4
AI in Health Care

◄◆————————◆►

Artificial intelligence in health care is perhaps one of the most significant illustrations of technology's capacity to revolutionize industries. At the heart of this transformation is a commendable aspiration: to enhance patient outcomes, boost accessibility, and improve the efficiency of healthcare systems worldwide.

The impact of AI on disease diagnosis is already profound, with machine learning algorithms capable of identifying complex patterns in medical images, such as CT scans or MRIs, which might elude even the most experienced clinicians. These AI systems offer the potential for earlier detection of life-threatening conditions like cancer or cardiac disease, thus increasing survival rates and reducing the costs associated with late-stage treatment.

AI also has the potential to transform patient care, particularly with the rise of telemedicine and remote patient monitoring. AI-enabled virtual health assistants can provide round-the-clock support, answer patient inquiries, provide reminders for medication, and even monitor patient health parameters. Such AI applications could reduce the burden on healthcare providers and facilitate better patient adherence to treatment plans.

Yet amidst these promising advancements, several significant ethical and practical challenges arise. One of the most pressing is data privacy. With AI requiring vast amounts of patient data to function effectively, the risk of data breaches or misuse of personal health information is a genuine concern.

Furthermore, in the event an AI system makes a medical error, the issue of accountability also arises. Who bears responsibility in such cases—the health care provider, the AI developer, or the AI system itself? Resolving these questions is imperative for the safe and fair use of AI in health care.

Lastly, relying too much on AI in health care may diminish the importance of the human touch, which is a fundamental part of patient care. Balancing technology and human interaction are vital in the development and implementation of AI in health care.

AI has the capability to usher in a new era of health care—more efficient, precise, and personalized. Yet we must navigate the accompanying challenges with a focus on patient welfare, ethical standards, and fair access to these advancements. Only then can we fully harness the transformative power of AI in healthcare.

Essay 5
VR Beyond Games

◄◆────────────────────◆►

I n the public's mind, virtual reality (VR) might be synonymous with immersive gaming experiences. However, VR extends far beyond the latest quest in a digital landscape. The education sector is being revolutionized by VR, offering new avenues for interaction and expanding our potentialities like never before.

In the sphere of education, VR is poised to revolutionize learning. Instead of students passively reading about the Roman Empire or molecular biology, they can explore these worlds directly through VR. This method can facilitate a deeper understanding and retention of information, making education an active rather than passive process. It is an exciting vista where the classroom has no boundaries, and experiential learning is accessible to all, regardless of geographical location.

Health care, too, is understanding the power of VR. Surgeons can now practice complex procedures using VR before they perform actual operations, enhancing their skills and reducing patient risk. Similarly, VR is used in physical and psychological therapy, helping patients recover from injuries or cope with mental health issues in safe, controlled virtual environments.

In the architecture and real estate sectors, VR allows for realistic, 3D walkthroughs of buildings and homes that are yet to be constructed. This technology not only helps architects visualize their projects more effectively but also allows buyers to experience a property as if they were physically present, thus aiding their decision-making process.

While these advancements paint a bright picture for the future of VR, we must also confront a handful of challenges. The high cost of VR technology may limit its accessibility, deepening the automated divide. As with any automation, there are health impacts and considerations to consider, particularly relating to prolonged use and the probable addiction to VR.

In closing, we find that VR's journey through games is just beginning. The uses are transformative and countless, suggesting a future where our relationship with the electronic world becomes more seamless. It is important to make sure that VR usage is ethical, reasonable, and designed for the good of humanity as we move forward.

Essay 6
VR and Social Interaction

Often, when we think about Virtual Reality (VR), the image that arises is solitary—a person disconnected from their surroundings, immersed in a world of their own. However, this visual does not encapsulate the entire story. The actual power of VR might not be its ability to isolate but to connect, offering new arenas for social interplay that hold both enormous potential and complex challenges.

Through VR, we can build digital communities unconstrained by geography. These spaces allow users to interact in environments that are inaccessible in the physical world. The implications are especially profound for those who are disabled or housebound. With VR, they could attend concerts, meet friends in exotic locations, or even walk again, if only virtually.

Moreover, VR's capacity in professional settings is notable. Remote work has become increasingly prevalent, and VR could transform it further. Imagine an immersive digital workspace where isolated workers collaborate as though they were in the same room—brainstorming on an automated whiteboard or exploring 3D models of the latest product design. This could enhance productivity while mitigating the isolation often associated with remote work.

Yet as we celebrate the capabilities of VR in communal interactions, we should not overlook the emerging problems. The question of authenticity arises—can a virtual hug replace a real one? While VR can create a sense of presence, there are aspects of physical interaction that it cannot replicate—at least not yet.

There is a risk that VR could become a substitute rather than a supplement for face-to-face contact. The possibility of "VR addiction" and its impact on mental health needs careful exploration and monitoring. It is also essential to understand that while VR can connect us across continents, it could also exacerbate the digital divide between those who have access to such technology and those who do not.

The effects of VR on public interaction are only beginning to unfold. While there are difficulties, the potential of VR to enrich our lives is enormous. We stand on the cliff of a new frontier in human interaction, where we must balance technological advancements with the preservation of the human aspects of connection.

Essay 7
VR and Mental Health:
Potential and Pitfalls

◀◆──────────────◆▶

Virtual Reality (VR) opens up vast new frontiers in mental health treatment. Its immersive qualities allow us to recreate, understand, and therapeutically manipulate the complex cognitive and emotional states that underlie mental health disorders.

One of the most promising applications of VR is in exposure therapy, especially for individuals dealing with phobias or post-traumatic stress disorder (PTSD). By recreating triggering situations in a controlled environment, VR allows therapists to expose their patients to fear-inducing stimuli, fostering resilience over time. The world within the VR headset is simultaneously genuine enough to elicit genuine reactions and artificial enough to provide a safety net.

In addition to treatment, VR holds the potential to train emotional regulation skills. Stress management techniques such as mindfulness can be taught in immersive environments, increasing their effectiveness. Picture a patient with anxiety learning deep breathing exercises on a calm, virtual beach. Their respiration rate and heart rate are monitored and integrated into the VR experience to provide real-time feedback.

But as we venture into this brave new world of digital therapy, we must also be cognizant of the pitfalls. As much as VR can help treat mental health conditions, it could, if used irresponsibly, contribute to them. We must know the risk of overuse or even addiction to VR environments. While VR can facilitate healing, it could also potentially be used to inflict psychological harm. We need to grapple with these ethical implications to avoid the adverse effects of VR technology on mental health.

Essay 8
VR in Future Workplaces:
Prospects and Challenges

Virtual reality (VR), once the purview of science fiction, is increasingly becoming an interesting facet of our everyday reality. Beyond its well-established role in gaming and entertainment, VR holds considerable potential to revolutionize the modern workplace.

In an age where remote work is more common, VR can replicate and even enhance the experience of physical presence. Imagine donning a VR headset and stepping into a meeting room halfway around the world, interacting with colleagues as if in person. Or perhaps using VR to recreate the serendipitous encounters of an office environment—those chance meetings at the coffee machine—can spark innovation.

Another fascinating application of VR lies in the realm of training. For instance, medical students could perform surgeries in computer-generated environments, making mistakes without dire consequences, before progressing to real-life patients. Firefighters and engineers could utilize virtual simulations to practice extinguishing fires and managing nuclear reactors.

As we envision these exciting prospects, we must not overlook the trials that accompany them. One of the central concerns is the effect of VR on the human psyche. Prolonged exposure to simulated worlds may impact our sense of reality with outcomes that are not yet comprehensible. Could the blurring of the real and virtual lead to cognitive dissonance or even psychological disorders? The impact of extended VR use on physical health, particularly vision and posture, is another critical aspect that warrants attention.

Second, privacy and data security come to the fore. If our workstations become virtual, what happens to the security of sensitive information? As our interactions with colleagues move to VR platforms, we need robust systems to protect our secrecy and ensure data integrity.

Third, we must contemplate the risk of a widening digital divide. As VR technology becomes integral to organizations, those who lack access to it are at risk of being left behind, exacerbating existing inequalities.

Virtual reality stands poised to reshape our systems in ways we are only beginning to comprehend. It promises increased flexibility, enhanced collaboration, and immersive training experiences. However, the problems it presents are substantial and cannot be dismissed lightly. As we forge ahead into this novel territory, it is imperative to navigate these waters with foresight and care, ensuring that our future workplaces are inclusive, safe, and conducive to human flourishing.

Societal implications: privacy and cybersecurity

In our digital age, the boundary between public and private realms is increasingly fluid. The ease of access to information and the linked nature of our lives have raised new concerns, particularly around privacy and cybersecurity.

The ubiquitous nature of the Internet has revolutionized how we communicate, how we work, and how we access information. The trade-off, however, is that our digital footprints—traces of our identities, behaviors, preferences, and activities—are now scattered across the digital landscape, exposed to the gaze of corporations and governments.

In this age of big data, these fragments of our digital selves are more than mere information; they are valuable commodities. Companies harvest and process this data to predict our behaviors, tailor our experiences, and even influence our decisions. Governments use them to enhance public services, maintain security, and, sometimes, exercise control. As our lives become increasingly digitized, the question arises: where does our right to privacy stand?

Simultaneously, we are also facing escalating threats in cybersecurity. As we depend more on digital networks, from our banking systems to our power grids, we become more vulnerable to cyberattacks. Data breaches, identity theft, and ransomware attacks—these are no longer just the stuff of fiction but realities that millions face.

Our increasing reliance on interconnected systems has elevated the risk and potential impact of cyberattacks. A single breach can lead to widespread disruptions, affecting individuals, corporations, and nations alike. Hence, the significance of robust cybersecurity measures cannot be overstated.

Yet it is not just about technological solutions. We must also contend with the societal and ethical dimensions of these encounters. In our pursuit of security, how much personal seclusion are we willing to sacrifice? How do we ensure that our responses to cyber threats do not infringe on our rights and freedoms?

In the following essays, we will dig deeper into these issues. We will explore the complexities of maintaining privacy in an increasingly interconnected world, the challenges and strategies for ensuring cybersecurity, and the ethical dilemmas we face in navigating this delicate balance. In doing so, we aim to foster a deeper understanding of these pressing concerns and encourage thoughtful discussion on how we can address them.

Essay 9

Privacy in the Age of Big Data:
An Uncharted Frontier

T he advent of the digital age, specifically the era of big data, presents a complex conundrum in our pursuit of privacy. Every click, every swipe, and every online interaction leaves a trail of information—breadcrumbs scattered in the vast wilderness of cyberspace.

This constant creation of specifications, seemingly harmless, adds to a complex web-based depiction of our habits, preferences, and behaviors. Enterprises, governments, and research institutions use these profiles to shape our experiences, formulate policies, and even predict our future actions.

The benefits of big data are undeniable. Its application has propelled advancements in many fields, from healthcare, where predictive models help identify disease patterns, to urban planning, where traffic details inform infrastructure development. However, amidst these promising developments, the issues of concealment and consent lurk in the background, sometimes overlooked but never truly distant.

In the age of automation, there is a fundamental paradox to confidentiality. Despite valuing this, we voluntarily reveal significant amounts of private material electronically. Is this a contradiction or a new norm for the virtual citizenry?

Second, the legal framework for online privacy is still in a state of flux. Different countries have their own set of regulations, and the internet, as a global entity, does not respect national borders. How, then, do we ensure the protection of privacy in this intricate, intertwined digital world?

A third concern arises from the corporate handling of information. Corporations like Google and Facebook have faced scrutiny over their documentation practices. But beyond these tech giants, a plethora of smaller entities also accumulate and trade user records, often without explicit consent. Is it then just to place the onus of privacy protection solely on the user?

As we explore this unknown area, it is essential to aim for a middle ground that allows the use of big data while safeguarding personal space. This will require a multifaceted approach—legal measures, technological solutions, and a cultural shift in our understanding of personal data protection.

Safeguarding confidential information in the age of big data is a colossal task that demands constant vigilance, innovative strategies, and a steadfast commitment to ethical facts protocols. As we shape the computerized landscapes of the future, let us ensure that the right to privacy remains a cornerstone, a beacon guiding our progress.

Essay 10

Surveillance Technologies and Society:
A Careful Balancing Act

The struggle between security and liberty has consistently existed throughout history, but it has never been more apparent than in our modern society, which is characterized by rapid technological advancement. At the forefront of this dynamic is the ever-developing field of surveillance technology, which, while promising increased safety, poses profound questions about our societal values, particularly privacy and individual freedom.

Monitoring tools, including facial recognition software, data mining algorithms, and drone technologies, are now ubiquitous. These tools promise enhanced safety measures, from counterterrorism operations to community protection protocols. The tangible benefits are irrefutable—swift identification of threats, efficient law enforcement, and perhaps a general deterrence for unlawful activities.

Yet one must question, at what cost do these benefits come? And who pays the price?

The character of observation technology involves a built-in violation of individual territories, both physical and digital. This intrusion raises valid concerns about privacy, autonomy, and the potential misuse of such devices. To put it simply, in our bid for a safer community, are we inadvertently creating a surveillance state where Big Brother ceaselessly watches?

The question of authority over these machines further complicates the discourse. While governments wield these tools under the banner of national safety, the private sector's burgeoning influence in this domain cannot be overlooked. The collection, storage, and analysis of data by corporate entities, under the guise of "improving user experience," presents another layer of security, often less scrutinized than state actions.

The unequal application of monitoring technologies, more acute in marginalized communities, provokes discussions about social justice. Are these tools exacerbating societal divisions and perpetuating biases?

As we deal with these questions, we should keep in mind that technology reflects the intentions of its users and is simply a tool. Therefore, to tackle the issues posed by monitoring equipment, a multifaceted strategy is needed—revising legal frameworks, cultivating a sense of responsibility, and encouraging openness in such use.

In the end, observation mechanisms and society are taking part in a sensitive interplay, a counterpoise between protection and autonomy. The path forward is not to shun these innovations but to navigate their complexities judiciously, ensuring that our pursuit of security does not trample on our treasured freedoms.

Essay 11

Balancing National Security and Privacy:
Striking the Delicate Equilibrium

◄━━━━━━━━━━━━━━━━━━━━━━►

When contemplating the landscape of the twenty-first century, we confront an intricate paradox where a taut thread suspends between two monumental pillars: national security and personal privacy. The challenge of our time is not to lean toward one or the other but to navigate the nuanced equilibrium that preserves both.

National security has grown more complex with advancements in technology and global interconnectedness. The emergence of nebulous threats, from cyberterrorism to sophisticated espionage, justifies the need for advanced surveillance systems and intelligence-gathering mechanisms. The preservation of a nation's sovereignty and the safety of its citizens rightly holds paramount importance in any government's mandate.

In contrast, the basic element of personal secrecy is interlaced into the composition of democratic governance and individual independence. It is not a luxury but a fundamental human right. Yet, in the modern digital landscape, the erosion of this is increasingly prevalent, often under the shadow of enhanced security measures.

The tension between these two facets is not a zero-sum game, despite how frequently it is portrayed. Indeed, it is a complex interplay that requires astute navigation and persistent negotiation. The task ahead, therefore, is not to diminish one in favor of the other but to understand how they can coexist in harmony.

Legislative measures play a critical role in this endeavor. Laws must adapt to the fluidity of technological advancements, protecting citizens' privacy rights while enabling necessary security measures. The oversight and regulation of surveillance programs are imperative to preventing potential overreach and misuse.

Meanwhile, the adoption of privacy-preserving technologies, such as end-to-end encryption and anonymization techniques, can create a buffer, allowing security operations to coexist with private digital spaces. Such tools epitomize the potential symbiosis between security and privacy.

Finally, fostering an open dialogue between policymakers, technologists, and the public can engender an environment of understanding and consensus. Transparency in governmental actions and clarity in the confidentiality policies of automated platforms help users comprehend and navigate the security-privacy nexus.

The pursuit of national security should not cause the forfeiture of privacy, just as the protection of privacy should not leave our nations vulnerable. Striking a balance is not only desirable; it is absolutely necessary to preserve the societal values we cherish.

After all, in the dance between security and privacy, music is a democracy, and we must ensure it never stops playing.

Essay 12

The Role of AI in Cybersecurity:
An Emergent Guardian

s the complexity of our digital existence deepens and the danger of cyber hazards increases, a new sentinel has arisen in the realm of cybersecurity—artificial intelligence (AI). As a complex tapestry of machine learning, predictive analytics, and pattern recognition, AI serves as a critical linchpin in the defense against a constantly changing suite of cyber threats.

Cybersecurity, in its essence, is an exercise in anomaly detection. Conventional methods struggle with the deluge of data and the increasingly sophisticated nature of attacks. This is where AI, with its ability to sift through large volumes of data and discern patterns in nanoseconds, holds the probability of being transformative.

AI-powered cybersecurity systems can harness machine learning to adapt and develop, learning from each interaction and sharpening their predictive capabilities. Such techniques can identify possible menaces and vulnerabilities, often before they are exploited, and execute rapid, automated responses. The advantage lies not just in detection but also in the alacrity of response, an element often decisive in the world of cybersecurity.

However, it is important to acknowledge that while AI can be a potent ally in cybersecurity, malicious entities can also weaponize it. Adversarial attacks designed to deceive AI systems, the creation of sophisticated phishing algorithms, and the use of AI for perpetrating deep-fake scams are but a few examples.

Navigating this double-edged sword calls for an approach that is both technologically astute and ethically guided. To thwart the misuse of AI in cybersecurity, we must accompany its deployment with robust protection mechanisms. This includes rigorous testing against adversarial strikes, continuous monitoring of AI behaviors, and transparent reporting of AI's decision-making processes.

We must also consider the implications of entrusting such a critical domain to artificial intelligence. While AI can provide immense technical prowess, the responsibility for cybersecurity cannot be fully abdicated to machines. The human element, with its capacity for ethical judgment, intuition, and nuanced understanding, must remain integral to the process.

Looking ahead, we can envision a cybersecurity landscape where AI and humans work synergistically. Machines give rapid computational might and relentless vigilance, while humans bring the ability to comprehend context, make value-based judgments, and provide creative problem-solving.

Artificial intelligence, in its role as an emergent guardian of cybersecurity, embodies a fusion of opportunity and challenge, of immense potential and noteworthy risks. As we chart the path forward, the key lies in harnessing this powerful tool judiciously, responsibly, and with a clear understanding of the multifaceted implications of its use.

Essay 13
Social Media's Role in Modern Democracy

D emocracy, a system that stands on the pillars of freedom, equality, and transparency, finds itself in a paradoxical tango with social media—a tool that can both illuminate and obfuscate these very ideals. The inextricable fusion of these two entities has given birth to a new form of participatory politics where the individual is not an observer but an active contributor.

Initially, perhaps social media, with its advent, extended the promise of greater democratization. It offered a platform for dialogue, expanded the breadth and depth of information accessible to the public, and toppled down the barriers of geographical distance. It gave power to the people—to have a voice, to deliberate, to dissent, and to decide.

Notable movements, such as the #MeToo, Arab Spring, and the Black Lives Matter campaigns, provide vivid testament to the potency of social media in galvanizing collective action. Social media platforms served as a nexus, a general square where dialogue flowed freely and where solidarity crystallized into action.

Yet we must also recognize the other side of the coin, where social media serves as a double-edged sword in our democratic discourse. Amidst the cacophony of voices, the authenticity and accuracy of information are increasingly shrouded in ambiguity. The susceptibility of social media programs to the propagation of misinformation, deepfake technology, and divisive content holds the potential to undermine the autonomous fabric by stirring civil discord, manipulating public opinion, and eroding trust in institutions.

In addition, the algorithmic nature of these platforms, designed to keep user attention, creates echo chambers that reinforce existing beliefs while muffling dissenting views. Instead of fostering an informed citizenry, this inadvertent effect breeds polarization— a stark divide that is antithetical to the consensus-building endeavor that democracy embodies.

It is this complex duality of social media's function in a modern self-government that requires a nuanced understanding and measured approach. As society navigates this uncharted terrain, it becomes essential to strike a delicate balance between leveraging the potential of social media and mitigating its perils. This balance needs judicious regulatory oversight, digital literacy education, and the ongoing commitment of tech companies to prioritize societal good over algorithmic engagement.

Social media is neither inherently democratic nor undemocratic; it merely reflects and amplifies the existing structures and behaviors in society. Therefore, the question is not about the role of social media but how we choose to wield this tool in our processes. The answer to this question will shape not only the future of our democracies but also the nature of our human interconnectedness in this digital age.

Essay 14
Emerging Models of Democratic Governance

◄●─────────────────────●►

Democratic governance, a living and dynamic system, perpetually grows in response to societal transformations and global challenges. We are seeing new patterns that integrate traditional methods with modern technologies, innovation, and citizen engagement. These developing models offer promising prospects for enhancing the legitimacy, efficiency, and inclusivity of constitutional processes.

Direct digital democracy, or e-democracy, is an archetype of such a model, and it leverages technology to improve the channels for citizen participation. With e-voting, online petitions, and online platforms for governmental consultation and policy discussions, the automated domain is developing into a contemporary agora. It carries the potential to revive free assistance and engage and amplify the voices of those who were previously disengaged and silenced.

However, as we further digitize our self-governing processes, the need to address access disparities becomes paramount. Unless this digital divide is bridged, e-democracy could inadvertently lead to the marginalization of certain socioeconomic groups, thereby perpetuating existing inequalities.

Another rising example is participatory budgeting, a process where citizens partake in decision-making about communal expenditures. By involving the public in budget allocation, it promotes transparency, accountability, and fosters a sense of civic responsibility. This grassroots approach empowers people, leading to a better alignment of budget decisions with community needs and thereby cultivating public trust in politics.

However, involved processes are resource and time-intensive and, without proper safeguards, could be susceptible to capture by special interest groups. Therefore, ensuring broad, fair support and managing potential conflicts of interest is vital.

The model of deliberative equality has been gaining traction. It entails creating spaces for informed discussion among settlers and encouraging mutual understanding, respect, and compromise. Practices such as citizens' assemblies and polls are expressions of this model. They seek to transcend the superficiality of slogan-dominated politics and cultivate a deep freedom rooted in thoughtful social deliberation.

These forums can be resource-intensive, and their outcomes are often advisory rather than binding. Therefore, translating the insights from these deliberations into policy action remains a formidable challenge.

In conclusion, these emerging models of autonomous administration reflect a move toward more inclusive, participatory, and deliberative practices. They each offer unique advantages, yet they also come with their own set of challenges that must be managed. As we venture into this brave new world of experimentation, the focus should be on fostering a culture of continuous learning, adaptation, and innovation. It is in this spirit of openness and adaptability that autonomy, in its true essence, will thrive.

Essay 15
Human Rights in the Digital Age

In the modern digital age, our understanding and application of human rights are continually challenged and redefined. The online revolution has resulted in profound societal transformations, providing unprecedented opportunities for human development and social connectivity. It also presents an array of intricate dilemmas concerning human privileges that must be carefully examined and addressed.

Automated platforms have emerged as vibrant public spheres, enabling the free exchange of ideas and amplifying the reach of voices advocating for rights and freedoms. They have mobilized social justice movements, facilitated international solidarity, and held powers accountable. They have also been the battlegrounds of disinformation, hate speech, and online harassment, posing grave challenges to freedom of expression and opinion.

The widespread adoption of modern technologies has brought about a new frontier in privacy claims. Personal data, in this computerized era, has become a highly prized commodity, traded, processed, and analyzed for various commercial and governmental purposes. While this data-driven approach can yield considerable benefits, such as personalized services and evidence-based policymaking, it can also lead to intrusive surveillance, profiling, and manipulation. Thus, the right to privacy in this era requires rigorous safeguards to prevent abuse and misuse of personal data.

In this regard, encryption technologies and data protection statutes, such as the European Union's General Data Protection Regulation (GDPR), offer crucial bulwarks. Their effectiveness hinges on robust enforcement and the willingness of web-based corporations to prioritize users' confidentiality rights.

Digital technologies have also revolutionized access to information and knowledge, arguably expanding the right to education. Digital learning platforms can democratize education, making it accessible to communities traditionally disadvantaged because of geographical, economic, or physical barriers. However, this shift also stresses the technology disparity—the disproportion in access to technological infrastructure and digital literacy skills. Without addressing this divide, the transformation risks exacerbating existing disparities in the right's realization of education.

Finally, the digital age brings forth new encounters with the right to work and just working conditions. Automation and artificial intelligence are transforming the nature of work, creating novel opportunities but also risks. Critical questions are being raised about job displacement, workers' liberty in the gig economy, and the need for new skills and lifelong learning.

The digital age has implications for every aspect of human rights, requiring a rethinking of traditional frameworks. It calls for rights-based approaches that adapt to the new realities while upholding the fundamental principles of universality, indivisibility, and inalienability of constitutional rights. A digitally inclusive, proper, and rights-respecting world is not a given—it is a goal for which we must actively strive.

Global Challenges: Climate Change and Biotechnology

The specter of climate change looms over the global narrative, casting long shadows on our collective consciousness. Meanwhile, the rise of biotechnology stands as a double-edged sword—a beacon of promise and a source of trepidation. Both of these forces pose unprecedented ordeals for our times, prompting us to reevaluate our relationship with the environment and our very essence as human beings.

Climate change, the most pressing issue of our era, confronts us with an array of cascading effects—rising temperatures, melting ice caps, shifting rainfall patterns, and more frequent extreme weather events, among others. Yet the climate crisis is not merely an environmental issue; it is a crisis that touches on every aspect of human existence—from our health, food security, and economic structures to migration patterns and social inequalities.

To engage with climate change means grappling with complexity and uncertainty and challenging the paradigms of consumption and growth that underpin modern societies. Our response to this crisis will dictate the legacy we leave for future generations. It calls for transformative changes—in our technologies, policies, lifestyles, and mindsets.

Meanwhile, in the realm of biology, we find ourselves at the dawn of a new era. Biotechnology, the manipulation of living organisms to produce useful products, has shown remarkable potential. It has revolutionized medicine, agriculture, and environmental conservation, offering solutions to some of our most persistent problems. Yet, like all powerful tools, biotechnology carries risks and ethical quandaries.

The ability to manipulate the very fabric of life raises profound questions. How far should we go in altering our biology or that of other species? How do we ensure that the benefits of biotechnology are equitably distributed, avoiding exacerbating existing disparities?

The following essays will provide a more in-depth analysis of these global obstacles. We will explore the multifaceted impacts of climate change, the potential and pitfalls of biotechnology, and the societal and ethical considerations that arise. We hope to illuminate the complexity of these issues and provoke thoughtful discourse on how we can navigate these uncharted waters. The trials are immense, but so are the opportunities for innovation, adaptation, and growth.

Essay 16
Human Impact on Climate

As we contemplate our existence on this pale blue dot of a planet, we must confront our impact on its climate. Indeed, the natural world dances to the tune of our endeavors, a grim waltz orchestrated by anthropogenic activities. The scale of our influence on the planet's climatic systems is such that our epoch may well be remembered as the Anthropocene—the age of humans.

Foremost among our atmospheric trespasses is the rapid and continuing escalation of greenhouse gas emissions. The carbon we exhale into the atmosphere, primarily from burning fossil fuels and deforestation, is effectively wrapping our planet in a thickening thermal blanket. A wealth of scientific evidence, such as rising surface temperatures and melting ice caps, clearly shows that the carbon-based chorus is to blame for global warming.

However, merely acknowledging the existence of global warming is to skim the surface of a much deeper, turbulent ocean. The rising temperatures set off a cascade of environmental changes with devastating implications. They amplify extreme global weather events, such as floods, hurricanes, and droughts, which have an impact on the water cycle. They cause polar ice to melt and sea levels to rise, threatening coastal communities and island nations. Their actions cause a wrecking effect on biodiversity by disrupting habitats and making species become extinct at a rapid pace.

Simultaneously, we need to consider another significant yet often overlooked aspect of our weather impact—the alteration of the earth's albedo, or reflectivity. Our activities, ranging from urbanization to soot deposition on snow and ice, have been altering this critical climatic parameter, probably amplifying the warming effect.

Moreover, we should also be cognizant of our indirect influences on the atmosphere, chiefly through altering the earth's biogeochemical cycles. Our agricultural practices, for instance, significantly disrupt the nitrogen and phosphorus cycles, leading to the release of potent greenhouse gases like nitrous oxide and altering cloud formation.

As we contemplate these manifold effects, a critical realization must arise. The human influence on our surroundings is not just an environmental issue but also a socioeconomic and moral one. Those who have contributed least to the problem, particularly communities in the developing world, can withstand the climatic disruptions. Climate change poses severe challenges to global development and social justice, calling for an unbiased approach to mitigation and adaptation.

In concluding this exposition, the pressing question remains: where do we go from here? While the scale of the conundrum may seem overwhelming, it is within our human capacity and responsibility to chart a different course. This entails a radical transformation in how we generate energy, how we manage land, how we produce and consume goods, and how we view our place within the natural world. As we tread into the future, we must carry the weight of this truth: our impact on the climate reflects our values and choices as a civilization.

Essay 17
Effects on Biodiversity and Global Health

◄◆──────────────◆►

I n the grand drama of life on earth, humans play the role of the dominant species, often acting as directors of fate and fortune for the rest of the biosphere. The culmination of our actions, especially in the last few centuries, has stirred a great tempest in the natural order, ushering in a wave of transformation that permeates both biodiversity and global health.

Biodiversity is more than a list of species; it is a complex web of life. Biodiversity is more than just a register of all living things; it is the embodiment of variety and the expression of life's resilience and creativity. However, human activity is causing an unprecedented biodiversity crisis. The onslaught of habitat loss, pollution, overexploitation, and climate change is speeding up species extinction rates to levels not seen since the last mass extinction event.

Yet one might wonder, why should we care about the fate of other kinds? Our concern for biodiversity is not solely rooted in a sense of moral responsibility towards other life forms. It is also a matter of self-interest. Biodiversity underpins the ecosystem services upon which we depend, from the pollination of crops to the purification of air and water. Losing biodiversity can destabilize ecosystems, reducing their resilience to disturbances and potentially triggering ecological cascades that could disrupt these vital services.

Moving on to global health, we find ourselves inextricably entwined with the condition of our planet. Emerging infectious diseases such as Zika, COVID-19, and Ebola remind us that our disruption of the natural world can have serious wellness implications. Many of these illnesses are zoonotic, originating in wildlife populations before spilling over into humans. As we continue to degrade habitats and increase our interactions with wildlife, we create more opportunities for such spillovers.

Climate change also has wide-ranging implications for global health. Rising temperatures and changing precipitation patterns affect the distribution and intensity of vector-borne ailments like malaria and dengue. The escalating severity and frequency of heat waves can cause heat stress and other heat-related illnesses. Changes in rainfall and temperature can also adversely influence agricultural productivity, leading to food insecurity and malnutrition.

As we grapple with these effects on biodiversity and global health, we must reckon with the complex, interconnected nature of the problems we face. They are not isolated phenomena but symptoms of something broader—a reflection of our relationship with the natural world. Addressing them calls for a systemic, integrated approach that goes beyond treating symptoms and strives to remedy the root causes. We must promote a worldwide culture of concern and regard for the biosphere to recognize the inherent worth of all existence and comprehend the intimate association between the well-being of our planet and ourselves.

Essay 18
Socioeconomic Impact and Sustainable Future

◄●──────────────●►

When we turn our gaze to the future, we find it tinted with a mix of uncertainty and possibility. The trajectory of our society and economy has never been more pivotal, influenced as they are by a gamut of forces, not least by our actions and decisions. At the core of this narrative lies the issue of sustainability—how can we overcome present obstacles to secure a thriving and equal fate for all?

To comprehend this quandary, it is crucial to recognize that the concept of sustainability extends beyond "being green." It is about fostering a stable balance between social, economic, and environmental considerations. It demands an understanding of the interplay between these spheres and how decisions in one area invariably affect the others.

Our economic system, traditionally focused on relentless growth and resource consumption, has ushered in an era of unprecedented wealth and technological advancement. But this

58

progress has come at a significant cost. Environmental degradation, income inequality, and societal fissures serve as stark reminders of the unsustainability of this model. Consequently, we stand on the cusp of a crucial juncture, one that compels us to rethink our approach to socioeconomic development.

A viable outlook will require a transition to a more circular economy, one that decouples economic growth from resource consumption. This model values resource efficiency, waste minimization, and the closing of material loops through recycling and reuse. It also opens avenues for new business models and job creation, reducing the communal impact of the transformation.

Inclusivity and societal equity also underpin a sustainable future. A society where opportunities and benefits are disproportionately distributed cannot be sustainable in the long run. We assure you that the shift toward sustainability is just and promotes social unity rather than marginalizing those already vulnerable.

Lastly, we must address the environmental aspect of sustainability. In our pursuit of a tenable future, we should aim to live within our planet's means, respecting ecological boundaries, preserving biodiversity, and mitigating our contributions to climate change.

Technology will undoubtedly be instrumental in driving this shift. From renewable energy and green infrastructure to digital platforms that enable a sharing economy, technological innovation can provide the tools we need to establish a durable future. However, technology alone cannot secure sustainability. It requires appropriate institutional structures, policies, and—most importantly—a societal mindset that values sustainability.

To conclude, the socioeconomic impact of our actions today will shape our tomorrow. To ensure a sustainable and resilient future, we must reorient our socioeconomic structures toward sustainability, strive for social equity, and respect our planet's ecological boundaries. It is a formidable task, but we can achieve it by working together, making informed decisions, and staying committed to maintainability. The future, after all, is not a destination but a path that we carve out with our actions and choices.

Essay 19
The Era of Genetic Engineering

◄────────────────►

As we venture into the uncharted territories of the twenty-first century, we carry with us a toolkit of unprecedented power—genetic engineering. This branch of biotechnology presents an opportunity to explore the essence of life, with the potential to transform medicine, agriculture, and even our understanding of what it means to be human.

Genetic engineering refers to the manipulation of an organism's genes. It is a concept that sounds simple in theory, yet its implications are profound. The ability to add, remove, or change genetic material within an organism could alter the course of life as we know it.

In the realm of medicine, the power of genetic engineering is immense. Gene therapies, such as CRISPR-Cas9, offer a glimmer of hope for curing hereditary disorders, eradicating inherited diseases, and enhancing human health and longevity. Imagine a world where conditions like cystic fibrosis or muscular dystrophy could be consigned to the annals of medical history. This is the future that genetic engineering tantalizingly promises.

In agriculture, this advancement could aid in addressing the demands posed by a burgeoning global population and evolving climate conditions. Genetically modified crops have the possibility of increasing agricultural productivity, improving nutritional value, and creating more resilient crops against pests, diseases, and harsh environmental conditions. Such advances could play a pivotal role in food security and sustainability.

Yet alongside these remarkable opportunities, the era of genetic engineering also presents complex ethical dilemmas. The notion of "designer babies," wherein parents might choose their offspring's physical and intellectual attributes, prompts profound questions about the nature of humanity, equality, and individual rights. Who gets to decide what traits are desirable, and who should have access to such technology?

Additionally, the ecological risks associated with genetically modified organisms (GMOs) cannot be dismissed lightly. The potential impact of GMOs on biodiversity, the emergence of "superweeds" resistant to pesticides, and the unintended effects on nontarget organisms must be carefully considered.

The era of genetic engineering is undoubtedly upon us, a potent testament to human ingenuity and curiosity. Yet as we harness this powerful technology, it is essential to navigate thoughtfully, ethically, and responsibly. It can be a tool for remarkable good, but it must be wielded with a deep respect for the intricacy of life and an unwavering commitment to fairness and equity. After all, with great power comes great responsibility—a truth that rings true as we step into this brave new world of inherited potential.

Essay 20
Ethical Considerations of Gene Editing

◄━━━●━━━━━━━━━━━━━━━━━━━━━━●━►

As we march forward into the era of genetic engineering, we carry not just the promise of scientific advancement but also the weight of ethical considerations in our hands. Gene editing, a technique that offers us the ability to alter the DNA of living organisms, is at the forefront of this moral discourse. The exciting scientific potential it harbors must be balanced against the upright concerns it elicits.

At the core of gene editing's virtuous debate is the fundamental question: just because we can, does that mean we should? While we have developed the technology to rewrite life's code, the wisdom to use this capability is something we shall cultivate. The power to manipulate the hereditary makeup of organisms, including our species, comes with serious responsibility.

Gene editing promises a future free of transmissible disorders and life-threatening diseases. A single cut in our DNA using tools like CRISPR could eliminate the transmitted mutations that cause diseases like Huntington's or sickle cell anemia. This is a remarkable promise. But it also opens the door to nontherapeutic applications, such as the enhancement of human physical or cognitive abilities. Such an application raises the specter of a new form of eugenics where the rich could likely enhance their offspring, exacerbating existing societal inequities.

The potential use of gene editing in human embryos, known as germline editing, presents further conundrums. Changes made to the germline would be passed on to forthcoming generations, leading to permanent alterations in the human chromosome pool. This throws up considerations of consent, as subsequent descendants cannot agree with the heritable changes being made.

Moreover, our understanding of the human genome is far from complete. A beneficial genetic modification could have unforeseen negative effects down the line, creating new illnesses and health issues.

In the environmental sphere, gene editing in agriculture and wildlife can affect ecosystems and biodiversity in unforeseeable ways. For example, gene drives aim to spread specific traits throughout a population rapidly, which could help eliminate disease vectors like mosquitoes. However, this meddling with nature could have unforeseen ripple effects, disrupting ecosystems and possibly causing new problems.

To navigate the minefield of gene editing, robust public discourse and inclusive policymaking are crucial. To ensure a variety of perspectives are considered, discussions should be transparent and informed, involving scientists, ethicists, policymakers, and the public. Regulation must be global, as the implications of this are not confined within geographical borders. The question of how we employ it is not just a scientific or medical one—it is an issue that belongs to all of humanity. The collective voice of the world must be heard in deciding how we employ this potent technology to shape our future.

Essay 21
Biotech's Role in Food Security and Agriculture

◀━━━━━━━━━━━━━━━━━━━━▶

I n the grand theater of life, biotechnology has taken center stage, promising solutions to some of the world's most pressing problems, one of which is food security. With a swelling global population and the accompanying rise in demand for food, agriculture faces an uphill task. Biotechnology, with its repertoire of genetic modification and precision breeding, offers powerful tools to navigate this arduous journey.

The power of biotechnology is rooted in its ability to enhance agricultural productivity and resilience. Engineering crop varieties that are nutrient-rich, pest-resistant, and drought-resistant can address the urgent need for more effective and sustainable farming. This encompasses the development of genetically modified organisms (GMOs) that can grow in suboptimal conditions, enabling us to use lands previously deemed unfit for agriculture.

Take the example of Bt cotton, a crop genetically engineered to produce a toxin that is lethal to certain pests. Implementing Bt cotton has significantly reduced the need for chemical pesticides, leading to safer farming practices and increased yield. Similarly, the development of golden rice fortified with beta-carotene to address vitamin A deficiency in regions where rice is a dietary staple demonstrates how biotechnology can enhance the nutritional value of crops.

Despite the scientific promise, the adoption of biotechnology in agriculture has been met with resistance. Critics argue that naturally reformed crops pose risks to biodiversity and ecosystem conditions. The dominance of a single changed crop variety might result in a loss of genetic diversity, rendering our food systems vulnerable to disease outbreaks.

Furthermore, the ethical implications of changing nature for our benefit have been debated. Issues around intellectual property rights and the corporatization of the global seed supply have been raised, highlighting the socioeconomic indications of wide-scale biotech adoption in agriculture.

Balancing these considerations, it becomes apparent that while biotechnology offers potential resolutions, it is not a silver bullet. Its role in securing our food future should be integrated with a range of approaches, such as promoting agricultural biodiversity, supporting small-scale farmers, and encouraging sustainable farming practices.

Adopting biotechnology in agriculture is not just a scientific decision; it is a societal one. It warrants a broad-based dialogue that includes the voices of farmers, consumers, policymakers, scientists, and ethicists. Such a comprehensive approach will help us utilize the potential of biotechnology safely and morally, boosting agricultural efficiency and food safety while protecting the planet's integrity.

Contemporary Human Concerns:
Mental Health and Income Inequality

As we turn our attention inward, we see two crises that are interconnected and that shape the human condition in our time: the increasing mental wellness crisis and the widening gap of income inequality. These twin challenges are intertwined in complex ways, each amplifying the effects of the other and both demanding our urgent attention.

Mental health, for too long relegated to the periphery of healthcare discussions, has stormed to the forefront with an insistent demand for recognition. The issue's alarming scope reveals an increase in depression, anxiety, and other psychological disorders, which our times' pervasive uncertainty exacerbates. Yet this dilemma is not just about numbers; it is an emergency of compassion, understanding, and societal attitudes.

Mental health is intimately linked to the fabric of our societies and to the ways in which we construct our lives and our identities. It is a personal issue, yet it is also social. The choices we make as a society—about our economic systems, our social safety nets, and our educational and health care policies—all shape our collective mental well-being.

As we explore mental health further, we come across income inequality in this discussion. The disparity in wealth and income has been a longstanding concern, yet the scale and intensity of this issue have escalated in recent years. The richest fractions of society continue to amass wealth at an unprecedented pace, while many others struggle to make ends meet.

Income inequality, however, is not simply about material affluence. It is about opportunities, about access to quality education and health care, and about the ability to lead a life of dignity and security. Moreover, income inequality has significant implications for psychological aspects, contributing to stress, anxiety, and feelings of disenfranchisement.

These modern human concerns will be explored in the coming essays, where we will analyze the intricate connections between mental health and income inequality. We will examine the societal and economic forces at play, illuminate the personal and communal consequences, and ponder solutions. These challenges are complicated and ingrained in our communal structures, yet they are not insurmountable. Addressing these issues requires collective will, innovative thinking, and, most importantly, empathy.

Essay 22
The Silent Epidemic of Mental Health

The pursuit of understanding our minds, the wellspring of our thoughts, emotions, and behavior, has been a quest threaded through human history. This journey has brought us face-to-face with a silent epidemic: mental health disorders. They are the unseen chains that bind the quiet storms that rage within the minds of countless individuals worldwide.

Mental health disorders, encompassing conditions such as depression, anxiety, schizophrenia, and bipolar disorder, are often shrouded in stigma and misunderstanding. This unfortunate scenario hinders early detection, treatment, and social support, further exacerbating the silent nature of this epidemic.

There is a pressing need to acknowledge that psychological well-being is a component of fitness that is just as consequential as physical wellness. This notion is encapsulated in the World Health Organization's definition of health, which enshrines mental health as a crucial aspect of overall healthiness.

Despite the high prevalence and impact of psychological health conditions, they remain marginalized in many welfare systems. This is reflected in inadequate funding for mental health services, a lack of integration into general health care, and the scarcity of mental health professionals. Such facts emphasize the institutional disregard, which is a fundamental aspect of this soundless outbreak.

The consequences of these untreated conditions ripple outward, affecting not just individuals but also families, communities, and societies. They exert a substantial toll on productivity and socioeconomic development. The challenge of mental well-being is causing immeasurable distress, resulting in the hushed spread of psychiatric conditions being one of the most significant human rights concerns of our era.

However, the silence around psychological welfare is gradually being broken. Movements promoting its awareness are gaining traction, encouraging conversations that challenge stigma and demand change. Digital technology is redefining it, enabling teletherapy and digital mental health resources to reach those who would otherwise remain untouched by traditional services.

But the journey is far from over. Addressing the silent, widespread epidemic of psychological health requires a global, unified effort. It involves the assimilation of mental health care into general health care, the promotion of mental health literacy, and the mobilization of societal and political will to prioritize psychological health. Compassion and understanding are the keys to illuminating the way out of the shadows for those ensnared by mental health disorders.

It is our collective responsibility to ensure that the voices of those grappling with mental health disorders are not relegated to a silent echo but are amplified and heard. Only then can we hope to transform this silence into a loud revolution for mental health justice.

Essay 23
Social Media and Mental Health

◆◆━━━━━━━━━━━━━━━━━━━━◆►

As we sail into the uncharted waters of the digital age, we are inundated with a constant barrage of information. The vehicle for this knowledge flood, the lynchpin ŏf our connected world, is social media. It has reshaped the societal fabric, giving rise to novel forms of communication, engagement, and interaction.

The tendrils of social platforms permeate almost every aspect of our lives, including our psychological condition. In this essay, we examine the complex relationship between social media and mental health, which can be both empowering and detrimental.

On the one hand, social media can function as a catalyst for positive change. It provides platforms for disseminating mental wellness information and awareness, combating stigma, and encouraging help-seeking behavior. Its power to connect individuals fosters communities of support. It enables people from all corners of the globe to share their experiences, engage in dialogue, and find solace in shared narratives.

However, the dark underbelly of this platform paints a different picture. The constant connectivity and the incessant stream of content can trigger feelings of inadequacy, anxiety, and depression. The curated perfection often depicted on these platforms belies the reality of human imperfection, cultivating a culture of comparison that can batter self-esteem and propagate negative body image.

Social media programs can become arenas for cyberbullying, with the veil of online anonymity enabling damaging behavior. The victims, typically adolescents, are left to grapple with the psychological aftermath, which can range from psychological distress to suicidal ideation.

The scale of online harassment and its implications for psychological well-being raise critical questions about the accountability and responsibility of social media platforms. These questions center on content moderation, user safety policies, and the role these platforms should play in addressing the mental health repercussions of their use.

Understanding the effects of social media on psychological health also causes a nuanced appreciation of the influence of the "digital diet." Similar to a nutritional diet, the quality, quantity, and timing of our online consumption can influence our intellectual well-being.

While the dialogue around social media and psychiatric health is regularly steeped in negativity, it is vital to remember the capability for positive change these principles possess. Harnessing these while mitigating their harmful effects demands comprehensive strategies encompassing regulation, education, and automated literacy.

The narrative of social media and mental health is complex, and the final chapters are yet to be written. What is clear, however, is the urgent need for rigorous research, informed policy, and societal engagement to guide our understanding and response to this modern phenomenon. The welfare of our automated generation depends on it.

Essay 24
Mental Health in the Workplace

————◆———————————◆————

A s the gears of our professional world turn, the ghostly specter of mental health issues hovers silently over the workplace. Mental well-being at work is not a luxury but a necessity, a fundamental component of employee welfare and organizational success. This chapter explores the complex interplay of mental condition and work, where personal vulnerabilities often intersect with professional obligations.

First, we must acknowledge the magnitude of the issue. Work-related stress, anxiety, and depression are not isolated phenomena but ubiquitous realities. They are the silent productivity killers, causing considerable human and financial costs. Experts estimate the projected global economic loss due to psychological conditions to be a staggering three trillion dollars by 2030. This projection underscores the deep societal and economic ramifications of ignoring psychological well-being in the workplace.

Workplaces can be both the cause and the remedy of mental health issues. On one side, chronic stress, long working hours, bullying, and a lack of job security can precipitate or exacerbate psychological health problems. The expectations of the "always-on" culture, coupled with the pressure to perform, can create an environment conducive to stress and burnout.

However, the workplace also has the possibility of being a place of healing. Well-structured employees, well-being programs, effective communication channels, and a supportive organizational culture can significantly improve mental health. The job environment can act as a platform for early identification and intervention, reducing the stigma around mental health and promoting help-seeking behavior.

Understanding and tackling mental health in the workplace requires multifaceted approaches. Employers have a pivotal role to play, from implementing mental health policies and providing resources for employees to ensuring a supportive and nondiscriminatory environment. Effective leadership can foster a culture of openness where discussing mental health is as normalized as discussing physical health.

Employees, too, have a part in this narrative. Self-care strategies, setting boundaries, seeking help when needed, and supporting colleagues are vital elements of maintaining mental well-being at work. The power of peer support and empathy in the workplace should not be underestimated.

Furthermore, mental health support in the workplace should be comprehensive, extending beyond interventions aimed at the individual to address structural and cultural factors. Such an approach would involve promoting work-life balance, managing workload, enhancing job control, and addressing issues like harassment and discrimination.

The matter of mental health in the workplace is complex and multifaceted, requiring engagement at multiple levels—individual, organizational, and societal. Addressing it is not just a matter of corporate responsibility but an imperative for sustainable, inclusive growth. Mental health should be an integral part of our discussions, policies, and practices as we move forward with the future of the industry. It is important to create workplaces that value and nurture the human spirit, not just for economic reasons but as a moral imperative.

Essay 25
The Future of Mental Health Treatment

◄◆━━━━━━━━━━━━━━━━◆►

Mental health treatment is entering a period of innovation and redefinition as we look toward uncharted territories. A potent amalgam of advancing technology, developing research, and shifting social attitudes is leading us into a new era where treatment may undergo transformative changes.

In the arena of medical advancement, the future of mental health therapy shows promise and perplexity in equal measure. Traditional psychiatric medication often presents a "one-size-fits-all" approach that overlooks the individual's unique biochemistry. However, the emergence of pharmacogenomics, which studies how genes affect a person's response to drugs, may allow for personalized medication regimens, increasing their efficacy and minimizing adverse effects.

Psychotherapy is also witnessing a sea change. Cognitive behavioral therapy, dialectical behavior therapy, and other evidence-based approaches continue to be refined and expanded upon. Integrating Eastern mindfulness practices into remedies represents a further broadening of our therapeutic toolkit. Digital platforms are revolutionizing the delivery of a cure, making it accessible to those in remote areas or those who prefer the anonymity of online interaction.

Emerging therapies are pushing the boundaries of our understanding. Psychedelic-assisted therapy, involving substances like psilocybin and MDMA, is reentering the scientific discourse after decades of cultural stigma and legal restrictions. Early clinical trials suggest the potential for treating conditions like PTSD, depression, and anxiety, although further research is needed to fully establish safety and efficacy.

Novel neurotechnological remedies, such as transcranial magnetic stimulation and deep brain stimulation, are also gaining ground. While such techniques are in their infancy and carry risks, they hold promise for conditions that are resistant to conventional treatments.

Amidst these scientific advances, however, we must not lose sight of the human element. Mental well-being is not the absence of illness but the presence of well-being. Therefore, holistic, integrative approaches that consider lifestyle, sociocultural context, and individual strengths will probably gain prominence. As our societal understanding of mental health evolves, we may witness a shift toward prevention and early intervention rather than medication alone.

Moreover, the future of mental health treatment must address systemic inequities that limit access to care. Technological advancements can only make a meaningful difference if they are accessible and do not perpetuate existing fitness disparities.

The journey toward the future of mental health treatment is teeming with possibilities and challenges. As we traverse this path, we carry a dual responsibility—to remain open to innovation and maintain our commitment to ethical, evidence-based care. In the domain of mental health, our pursuit is not merely scientific discovery but human understanding, compassion, and healing.

Essay 26
The Widening Gap Between Rich and Poor

In the dramatic narrative of economic inequality, the division between the haves and the have-nots has emerged as an insidious and persistent conflict. The mounting disparity between the rich and the poor, far from being a peripheral issue, is a fissure running through the heart of contemporary society.

The dynamics of wealth accumulation and income disparity tell a stark tale. Wealth generation, in an ideal world, should be a tide lifting all boats. However, the reality today is closer to a geyser, pooling at the top. A small segment of society is amassing affluence at an unprecedented rate, while a substantial portion continues to grapple with stagnating incomes and escalating living expenses.

The reasons for this disparity are manifold and complex, intertwining the threads of globalization, technology, taxation policies, and systemic bias. The globalization of labor markets, for instance, has often led to a "race to the bottom," with multinational companies seeking the lowest compensation, suppressing income growth for workers.

Advancements in technology have created a dual-edged sword. While the digital economy generates immense wealth for those at the forefront, it can also contribute to job displacement and wage disparity, with automation affecting lower-earning jobs.

Taxation policies, too, play a significant role. Regressive tax structures, where the tax burden is imposed more heavily on the poor than the rich, can exacerbate income inequality. Progressive taxation, where the rich are taxed at higher rates, can serve as a tool for wage redistribution if appropriately implemented.

The consequences of this growing chasm extend beyond individual hardship. Inequality can stifle economic growth, fuel social unrest, and undermine democratic institutions. Moreover, it perpetuates a cycle of disadvantage as those born into poverty face obstacles in accessing quality education, health care, and opportunities for upward mobility.

Addressing this widening gap requires a concerted, multipronged approach. There is a pressing need to revisit our tax structures and labor laws, ensuring that they promote fairness and economic mobility. Investments in education and skill development can also help equip individuals for the opportunities and challenges of the digital economy. Furthermore, cushioning those affected by economic transitions requires the establishment of strong social safety nets.

The broadening imbalance between the rich and the poor is a mirror reflecting our societal values and decisions. It prompts us to question the kind of society we wish to create—one defined by stark divides or one marked by shared prosperity. As we grapple with this issue, we would do well to remember that economic systems are not forces of nature but human constructs amenable to change. It is in our hands to bridge this gap and, in doing so, to build a fair and inclusive world.

Essay 27
Automation's Impact on Income Inequality

◄─────────────────────────►

The ceaseless march of technology has brought us into an age of automation, a time when machines, algorithms, and artificial intelligence are redefining the contours of work. The impact of this paradigm shift is wide-ranging and multifaceted, but one thread that merits special attention is automation's role in the narrative of income inequality.

At its core, computerization is a process of substitution. Machines and algorithms are introduced to perform tasks previously done by human labor. While this certainly increases efficiency and productivity, it also displaces human labor, which could widen the income inequality gap. The impact of this is not homogenous across the spectrum of workers. It affects routine-intensive jobs disproportionately, jobs that are often at the lower end of the wage scale. Whether it is the factory worker whose role has been taken over by a computerized assembly line or the data entry clerk whose job is made redundant by intelligent software, the individuals withstanding mechanization often find themselves on the precarious edges of the economic structure.

Automation reaps substantial benefits for the owners of capital—those who invest in, develop, and deploy these technologies. The cost savings and efficiency gains from this are considerable, boosting profits and, thereby, the earnings of the capital owners. Automation has caused a divide between those who benefit from it and those who do not, which is making the income inequality gap worse.

In addition, it can have a polarizing effect on the labor market. It could lead to an erosion of middle-skilled jobs while boosting demand for both low-skilled and high-skilled roles. This phenomenon, known as "job polarization," may cause a "hollowed-out" labor market with more jobs at the extremities of the skill (and consequently, wage) spectrum, fueling further income inequality.

Addressing automation's role in income inequality involves thoughtful, initiative-taking strategies. Upskilling and reskilling initiatives can help workers transition to new roles in an automated economy. Social protection schemes, including unemployment benefits and retraining programs, are also crucial to cushioning the impact on displaced workers. To mitigate the disparity effects of automation, implementing policies that guarantee a more reasonable distribution of its advantages, such as progressive taxation or profit-sharing schemes, can be beneficial.

As we move forward in the age of automation, we must acknowledge the possibility of it worsening income inequality. The onus is on us, as a society, to shape the narrative of automation to ensure it becomes a tool for broad-based prosperity rather than a catalyst for disparity. By putting people at the heart of our response to industrialization, we can strive to build an inclusive economy where technology serves as an enabler, not a divider.

Essay 28
Income Inequality and Health Disparities

A s we navigate through the socioeconomic landscapes of the twenty-first century, we encounter an intriguing yet alarming paradox. Technological progress, economic development, and medical advancements have made us more prosperous and healthier than ever before. Yet beneath this veneer of progress lies a stark dichotomy—income inequality and health disparities persist and, in many ways, are widening.

Income inequality is not just a question of fairness or social justice. It has far-reaching impacts, notably on health. Research has established a strong correlation between income levels and health outcomes. People with lower incomes often bear a higher burden of health problems, ranging from chronic diseases to mental health issues.

One of the primary reasons for this disparity is inequitable access to health care. Those who are economically advantaged have better access to quality medical care, preventive services, and early diagnosis, resulting in improved wellness outcomes. On the contrary, those with lower wages, in the absence of comprehensive public medical help or affordable health insurance, may defer or forego necessary health care, leading to poorer health outcomes.

Apart from the availability of health care, income inequality has indirect effects on health. There is a greater likelihood for individuals with fewer earnings to live in neighborhoods with limited access to healthy food, fewer opportunities for physical activity, and greater exposure to environmental toxins, all of which can negatively affect their health. Financial stress and insecurity, which are more prevalent among lower-income groups, also contribute to adverse health outcomes, including mental health issues.

These health disparities, in a cruel twist of irony, can perpetuate the cycle of income inequality. Poor health can limit educational attainment and job prospects, leading to lower lifetime earnings. This negative cycle of poverty is not just a personal tragedy; it also represents a loss of human potential on a societal level.

Addressing these disparities requires a multifaceted approach that goes beyond the medical care sector. It causes social policies aimed at reducing income inequality, including progressive taxation, a strong social safety net, and investments in education and affordable housing. Health policies must address affordable and quality medical assistance access and the social determinants of health.

In summary, the intersection of income inequality and health disparities is a complex issue that requires an integrated, multi-sectoral approach. It challenges us not only to address the symptoms of poor health but also to confront the underlying social structures that contribute to these disparities. As we work toward solutions, let us remember that wellness is not an individual responsibility; it is a collective endeavor, a reflection of the kind of society we aspire to be.

Essay 29
Policy Solutions to Combat Income Inequality

The issue of income inequality looms large in our society today. With a wealth gap that only appears to be growing, it has become vital for us to search for solutions that can help mitigate this global predicament.

One must approach this issue from various angles. To begin with, the tax system serves as a prominent tool for addressing wage disparity. A progressive taxation system, where the rich are taxed at a higher rate compared to the less affluent, can help distribute wealth more evenly. However, it is critical to execute this policy judiciously, ensuring it neither discourages economic productivity nor discourages investment.

Labor market policies also play a pivotal role in combating income inequality. A substantial increase in the minimum wage, coupled with strengthening collective bargaining rights, could go a long way toward ensuring that workers across all sectors are paid a fair wage that matches the cost of living. Gender and race wage gap issues ought to be confronted head-on, endorsing a commitment to equal pay for equal work.

The role of government in providing comprehensive and quality public services cannot be overstated. Investing in public education, for instance, ensures everyone has access to the basic tool of socioeconomic mobility, regardless of their financial background. Quality health care services, affordable for all, are equally essential, as this not only enhances the well-being of individuals but also promotes productivity and industrial output.

Access to affordable housing also needs to be a priority. Gentrification and rising property prices often lead to a situation where individuals with lower compensation are forced out of their neighborhoods, aggravating disparity. The promotion of affordable housing policies can provide families with the stability and security to live with dignity.

One must look beyond these traditional policy solutions and explore innovative approaches. The concept of a universal basic salary, which guarantees a certain amount of pay to every citizen regardless of their employment status, is becoming more popular. Such an approach could provide a safety net for the most vulnerable, ensuring they are not left behind in our transforming economic landscape.

Ultimately, comprehensive, targeted, and innovative policy solutions are necessary to combat income inequality. To foster a society where opportunity is not confined to the privileged few but is a right enjoyed by all, it is important to ensure the equitable sharing of the fruits of economic development. This is not an economic imperative but a moral one, affirming the inherent dignity and worth of every individual.

The Final Frontier: Space

A new frontier, space, looms before us as we gaze toward the vast and enigmatic cosmos. The infinite expanse of the universe presents us with opportunities for exploration and discovery that dwarf anything we have known before. Yet this grand endeavor is not just about advancing our understanding of the cosmos; it is equally about deepening our understanding of ourselves, our capabilities, and our responsibilities.

Space exploration is the ultimate test of our scientific and technological prowess. The ability to traverse the great void, to land on distant celestial bodies, and even to envision colonizing other planets speaks to the remarkable advancements we have made in science and technology. But this journey, for all its grandeur, presents us with deep questions—what are the ethical implications of venturing into space? How do we balance the drive for discovery with the need for conservation? What responsibilities do we have toward these otherworldly realms and toward the lifeforms we may encounter there?

As we venture further into the cosmos, we must also turn our gaze inward toward the earth that we call home. The technologies that propel us into space can address pressing challenges here on Earth, from climate change to energy scarcity. Exploring space, in this context, is not only about venturing to the stars but also about preserving the planet that supports us.

The expedition to space holds potent symbolic value. It exemplifies human curiosity, thirst for knowledge, and capacity for wonder. It is a beacon of hope and a reminder that we are capable of extraordinary feats. And perhaps it is a call for unity. The challenges of space exploration require international cooperation and a collective pooling of resources and expertise. In this shared endeavor, we find a powerful antidote to the divisions that often plague us on Earth.

In the ensuing discussions, we will delve deeper into the possibilities and challenges of space exploration. We will examine the technological advancements propelling this journey, ponder the ethical dilemmas it presents, and explore the potential benefits for our planet and our species. The journey into space, like all great endeavors, is fraught with uncertainties and risks.

Essay 30
The New Space Age and Mars Colonization

◄━●━━━━━━━━━━━━●►

As we stand on the precipice of the new space age, our gaze is drawn toward the red planet Mars, which emerges in our collective imagination as the next frontier for human exploration and perhaps colonization. It is our responsibility to examine the principal issues that will affect this venture.

The prospect of a Mars settlement is steeped in a mixture of scientific curiosity, technological triumph, and existential necessity. With Earth wrestling with climate change and overpopulation, Mars presents itself as a potential "Plan B." However, it is crucial to recognize that the journey to make Mars habitable is one fraught with immense challenges.

From a technological standpoint, the hurdles are substantial, though not insurmountable. We must develop more advanced propulsion technologies to reduce travel time, devise ways to shield astronauts from radiation during the voyage, and establish self-sustaining habitats that can provide water, food, and breathable air in the inhospitable Martian environment.

The drive toward Mars colonization also presents a wealth of scientific opportunities. The planet's geological and climatic history might hold keys to understanding the nature of planetary evolution and the possibility of extraterrestrial life. Mastering the technologies needed could also have spillover benefits on Earth, from advancements in recycling and energy production to materials science.

Yet beyond these practical considerations lies a thicket of ethical and political issues that demand our attention. Who will govern a Mars colony? How will we ensure that the exploitation of Mars resources does not mirror the historical injustices of Earth's colonial past? These questions invoke the need for international regulations and agreements that ensure the peaceful, fair, and sustainable use of extraterrestrial resources.

Further, we must consider the implications of colonizing Mars for our species' identity. If we are to become an interplanetary species, how will this shift affect our perceptions of nationality, race, or religion? How would the narrative of human history be altered? The answers to these thorough questions will determine our future in ways we can hardly imagine.

In sum, the colonization of Mars is more than a feat of engineering. Yet we must ensure that wisdom, not hubris, guides this bold endeavor in a spirit of shared destiny, not narrow self-interest. As we set sail for a new world, let us bring the best of humanity with us.

Essay 31
Ethical Considerations of Space Exploration

◀◆━━━━━━━━━━━━━━━━━━◆▶

Traversing the expanse of cosmic shores, our journey into space is not just an exploration of the cosmos but also an examination of our ethical and moral compass. As we explore the galaxy, we must confront humane considerations and shape a moral framework that fits our interstellar era.

The very act of galactic exploration, while signifying the triumph of human ingenuity, provokes significant upright debate. One cannot evade the question: should we, as a species, invest enormous resources into planetary expeditions when such funds could ease suffering and hardship on our home planet? This dilemma encapsulates the classic conundrum of immediate needs versus long-term vision. While it is essential to address terrestrial issues, it is crucial to invest in space exploration, which promises immense scientific, technological, and existential dividends.

Respect for extraterrestrial life and environments is another fundamental aspect of the said exploration. Any life forms, even microbial ones, found beyond Earth deserve our respect and protection. We must grapple with the proper implications of planetary protection, both protecting other worlds from contamination by Earth life and preserving our planet from possible extraterrestrial biohazards.

Then there is the issue of exploiting space resources. With the prospect of mining asteroids and other celestial bodies becoming a workable reality, the question arises: who owns space? Any rush to claim the bounty of the cosmos risks repeating the historical inequities of colonialism. For space travel, there should be a universal agreement that views space and its resources as the shared heritage of humanity, ensuring fair distribution and sustainable practices.

The considerations for human colonization of other planets are enormous. If we alter other planets to suit human needs—a process called terraforming—we must ponder the morality of such a grand act of ecological manipulation. The treatment of space settlers, their rights, governance, and social structures are matters of significant deliberation.

Finally, there is the broader, more philosophical, and ethical issue of the potential consequences of contacting advanced extraterrestrial civilizations. The outcomes are unpredictable, with possibilities ranging from peaceful coexistence and mutual enrichment to conflict and even the end of human civilization. This is perhaps the most profound virtuous consideration, demanding extraordinary caution and collective decision-making.

To reach the stars, we must also look inward at our ethical core. The path should not be a trail of technological milestones but a voyage that elevates our moral understanding. The stars may be silent, but our ethical debates around space exploration will continue to resonate, guiding us toward a future that celebrates not just where we are going but who we are becoming.

Essay 32
The Implications of Discovering Extraterrestrial Life

I n the grand cosmic theater, the question of extraterrestrial life has captivated human curiosity for centuries. Should we one day validate its existence, this discovery would unfurl profound implications for science, philosophy, and our collective self-conception, indeed reverberating across our thoughts and even society.

From a scientific perspective, the discovery of extraterrestrial life would expand the boundaries of biological science. It would deliver a categorical affirmation of the Drake equation, which seeks to estimate the potential number of civilizations with which we might communicate in our galaxy. Biology would enter a new epoch, tasked with studying the biochemistry, evolution, and ecology of extraterrestrial organisms. Further, this discovery would either affirm or repudiate the theory of panspermia—the hypothesis that life on Earth was "seeded" from space.

Philosophically, the existence of life beyond our planet would compel us to redefine our understanding of life's meaning and purpose. An existential humbling would follow, diminishing our "geocentric conceit" and reinforcing our cosmic insignificance. Our faith systems, deeply entrenched in earth-centric narratives, would face profound questions, and perhaps find themselves on the brink of an evolutionary leap or crisis, demanding reinterpretation of ancient texts and doctrines.

The social implications are far-reaching. Such a revelation could catalyze a new sense of global unity, or "earth identity," enhancing our perception of the collective human family. As the distinguished cosmologist Carl Sagan once remarked, the discovery of extraterrestrial life might serve as a "mirror" through which humanity could see itself from a new perspective.

We must consider the potential negative implications. The discovery could cause public distress, elicit fears of the unknown, or even incite conflict over how to interact with these extraterrestrial entities. The history of human contact with "the other," be they unfamiliar cultures or species, is fraught with exploitation and conflict—a dark past we should be mindful not to repeat on a cosmic scale.

The prospect of extraterrestrial intelligence raises even more significant issues. Should we attempt communication, and if so, what should we say? Who gets to represent Earth in this interstellar dialogue? And what if they are far more advanced? Would we face subjugation or extinction, or could we benefit from their wisdom?

In conclusion, the discovery of extraterrestrial existence would irrevocably alter humanity's place in the cosmos, opening a new chapter in our scientific understanding, philosophical discourse, and social evolution. This event would be a testament to our adventurous spirit and insatiable curiosity—a moment that forever changes our view of the skies above and life below.

Reflections on Current Themes Shaping Humanity

◄◆─────────────◆►

As we survey the vast canvas of our era, a few vibrant themes stand out, their hues defining the spectrum of contemporary human experience. These themes, encompassing the realms of technology, sociology, ethics, and environment, are intrinsically intertwined, reflecting the complexity characteristic of our shared global narrative.

In the field of technology, the advent and acceleration of Artificial Intelligence and genetic engineering have indelibly marked our epoch. The promise of AI, which crowns the digital revolution, raises both exciting possibilities for societal advancement and unsettling concerns about human relevance in a world of machines. Genetic engineering, on the other hand, brings forth the tantalizing prospect of defeating diseases and enhancing our biological capabilities. Yet it also raises profound concerns, challenging our notions of identity, equality, and the very essence of being human.

Societal and ethical themes are interwoven with these technological shifts. The digital divide and automation have further

106

exacerbated income inequality, a persistent blight on our collective conscience. The enduring struggle for social justice, spanning across the complexities of race, gender, and class, remains as critical today as it has ever been. Amidst these challenges, we must grapple with the ramifications of our scientific advancements and the shared responsibility we bear in ensuring their fair deployment.

The specter of climate change, arguably the most potent existential threat we face, throws into sharp relief our relationship with the environment. While we deal with the repercussions of our past mistakes, we are also considering a new connection with the universe through Mars colonization and space exploration. Both of these pursuits underscore the quintessential human traits of adaptability and exploration, even as they underscore the urgency of our environmental stewardship.

The theme of mental health, which is now widely accepted as a critical aspect of holistic well-being, is a testament to our growing understanding of human complexity. From the societal structures affecting our mental well-being to the role of our digital lives, mental health's prominence as a discourse is a sign of progress in our collective empathy and self-awareness.

Lastly, the continuing pursuit of extraterrestrial life symbolizes our timeless quest for understanding the grand cosmic narrative and our place within it. Our natural curiosity and sense of wonder continue to shape our philosophical and scientific perspectives, encouraging us to consider our shared existence and universal unity.

Our exploration of these themes and their implications is done with a sense of humility and cautious optimism. As we navigate these currents, we collectively shape our present and future, embarking on a journey that is both exciting and daunting, rewarding and challenging, humbling and empowering.

The interconnectedness of themes explored

The themes dissected within the preceding chapters, though disparate in nature, reveal, upon closer examination, a vast interconnected web. They are threads in the fabric of our collective existence, their harmonies and discords shaping the symphony of the human narrative.

Take, for instance, the symbiotic relationship between technological advancement and social inequality. While the strides in artificial intelligence and automation herald a future of unprecedented productivity and innovation, they also, paradoxically, exacerbate the divide between the socioeconomic classes. As jobs become automated, the gap between the rich and the poor widens, painting a sobering picture of prosperity coexisting with destitution.

Similarly, the intersection between psychological wellness and our digital lives illuminates another facet of this intricate tapestry. Social media, which is valued for its ability to connect people worldwide, has unintentionally intensified feelings of loneliness and inadequacy, highlighting the contradiction in our technological progress. The need to redefine success and productivity is urgent in the workplace because of the pressure to perform in a highly competitive environment, which worsens mental health issues.

The ripples of income inequality, further magnified by technological disparity, also touch upon the shores of health disparities. The chasm between the rich and the poor extends beyond financial resources, penetrating the realms of health and longevity. In an era defined by medical breakthroughs, the cruel irony is that access to such advancements often hinges on wealth, a poignant reflection of our societal structures.

Our exploration of space and the quest for extraterrestrial life are not isolated endeavors. They mirror our aspirations, our need for exploration, and the unquenchable human spirit of discovery. Yet they also spotlight our relationship with our home planet, a timely reminder amidst the throes of climate change and a call to stewardship even as we reach for the stars.

Then, there are considerations that straddle across various themes. From the ethics of gene editing to those of space colonization, our moral compasses are constantly tested. We are at a crossroads where scientific progress meets ethical obligation, compelling us to reconsider our values and plot the path to our communal destiny.

As we explore each theme more deeply, we become more aware of their interconnectedness, highlighting the importance of comprehensive solutions that take these interdependencies into account. For it is only through such understanding that we can hope to address these multifaceted challenges, shaping a future that respects the intricate balance between progress and equity, innovation and ethics, discovery and responsibility.

A call to action for addressing these challenges

Faced with these numerous challenges, it would be easy to succumb to a sense of overwhelming despondency. Yet history has repeatedly shown that humanity's greatest strength lies in our ability to adapt, innovate, and persevere in the face of adversity. We must rally this resilience once more and direct it toward comprehensive solutions.

First, we must work toward developing proper guidelines for technological advancement. The current pace of innovation far outstrips our ability to legislate its ethical implications. Robust ethical frameworks must keep pace with the increasing prevalence of gene editing and artificial intelligence technologies. This requires a concerted effort from scientists, ethicists, policymakers, and society at large to foster a dialogue that respects scientific innovation while prioritizing human dignity and equity.

In parallel, we must address the gap in social and economic inequality that threatens to deepen with advancing technology. Redistributing wealth and resources is not just a moral imperative but a pragmatic one. We must work toward progressive tax systems, invest in education and reskilling, and promote policies that foster social mobility and income equity. An inclusive approach will ensure that the fruits of our progress are shared, contributing to societal stability.

The silent epidemic of mental health requires our immediate attention. We must promote a cultural shift that values mental health alongside physical health, combat stigma, and provide accessible mental health resources. Encouraging responsible usage in the digital age requires platforms to incorporate features that prioritize users' mental health. Organizations must foster environments that value employee well-being, creating a culture that understands the pressures of the modern workplace and provides support mechanisms.

It is crucial that we approach the new era of space exploration with a sense of responsibility. The prospect of Mars colonization and the potential discovery of extraterrestrial life bring forth new ethical considerations. As we strive for the stars, we must respect all life forms and act responsibly as caretakers of the universe.

Most critically, our approach to these challenges must be holistic. The themes we have explored are intricately connected, a microcosm of the complex world we inhabit. Isolated solutions will not suffice in a world where the boundaries between technology, society, and individuals are increasingly blurred. We need comprehensive, interdisciplinary strategies that acknowledge this interconnectedness.

The challenges before us are monumental but not insurmountable. With a collective will, ethical responsibility, and an inclusive vision of progress, we can navigate this intricate tapestry, shaping a future where innovation thrives, equity is upheld, and human dignity is at the heart of our advancement. This is our call to action. Let us rise to meet it.

Concluding Remarks

When we look to the future of humanity, we see a horizon that is overflowing with complexity, opportunity, and paradox. We live in a time of unparalleled technological prowess that holds promises of solutions to problems that have dogged humanity for centuries. Yet these advancements, as wondrous as they are, also cast long shadows we must not ignore.

Throughout this exploration, it has become clear that humanity is intertwined in a dynamic dance with technology. From the contours of our mental health to the shape of our social structures to the potential colonization of other planets, technology plays a pivotal role. It is not an omnipotent conductor, but an instrument that we wield. And like all tools, its impact—positive or negative—reflects the hands that hold it.

We must remember that beneath the statistics of inequality, behind the screen of social media, and beyond the science of genetic engineering, there are human stories—narratives of individuals and communities navigating these turbulent times. This is the beating heart of our discourse—the human element. We must ensure that the whirlwind of progress does not sweep away the essence of our shared humanity.

Our journey through these topics has not been to predict the future with unerring precision, for such a feat is beyond the reach of even the most profound minds. Our aim was to shed light on the trends that impact our collective fate and encourage a more comprehensive dialogue. The road ahead is not a predestined path but a trail that we carve out with our actions, decisions, and, most critically, our values.

Our story is not yet written, and therein lies our greatest hope. We should play an active role in shaping our future rather than merely observing from the sidelines. By acknowledging and addressing the challenges that lie ahead and being conscious of the interconnectedness of our actions, we are taking the first steps toward steering the course of our shared destiny.

This is not a tale of inevitable doom or assured utopia, but a call to mindful, purposeful action. As we approach a significant transformation, let us make choices guided by both progress and human values of empathy, equity, and ethical stewardship. Let our progress not just be defined by how far we reach but also by how deeply we see, how widely we embrace, and how wisely we act.

As we close this chapter, remember that the most important part of this narrative is yet to be written. And the authors of that chapter are you and us—all of us together. As we step into the future, let us do so with awareness, with courage, and most importantly, with hope. Our shared aspirations and collective dreams will shape the destiny of humanity, forging a path toward a remarkable future.

Glossary

artificial intelligence (AI). The emulation of human intelligence processes by machines, particularly computer systems.

automation. The development and utilization of technologies to generate and distribute goods and services with limited human involvement.

biotechnology. The application of living systems and organisms in the development and production of products.

blockchain. A publicly available and time-ordered digital ledger for documenting transactions.

climate change. Prolonged shifts in global temperature, precipitation, wind patterns, and other climate metrics.

cybersecurity. Shielding systems, networks, and programs against digital intrusions.

data mining. The process of discovering patterns and knowledge from copious amounts of data.

data privacy. The aspect of information technology that deals with the ability of an individual or organization to control what digital data can be shared with third parties.

encryption. The method by which information is converted into a secret code that hides the information's true meaning.

extraterrestrial life. Life that may exist and originate outside the earth; the existence of which is still theoretical.

fossil fuels. Energy-rich resources generated by the decomposition of organic matter over geological time.

gene editing. Targeted modification of DNA at a specific location in an organism's genetic blueprint.

greenhouse gases. Atmospheric gases that retain heat, leading to the greenhouse effect and the escalation of global temperatures. Notable examples include carbon dioxide and methane.

globalization. The interconnectedness and integration of individuals, businesses, and governments across the globe.

hybrid vehicle. A dual-powered vehicle that employs both conventional internal combustion engine technology and electric propulsion systems.

income inequality. Disproportionate income distribution, highlighting the unequal apportionment of household or individual earnings within an economy.

job automation. The use of automated machinery to perform tasks traditionally done by humans.

knowledge economy. Denotes an economy that thrives on the quantity, quality, and accessibility of digital information, rather than conventional means of production.

machine learning (ML). Adaptive artificial intelligence, denoting a type of AI that can adapt and improve its performance based on experience and learning.

Mars colonization: The hypothetical human habitation and exploitation of the planet Mars.

mental health: A person's condition regarding their psychological and emotional well-being.

nanotechnology. Atomic and molecular manipulation, highlighting the precise control and manipulation of matter at the atomic and molecular scale.

orbit. Curved trajectory caused by gravitational attraction, highlighting the path of an object as it orbits around another body.

photonics. The physical science and application of light.

quantum computing. Harnessing the collective behavior of quantum states, such as superposition, interference, and entanglement, for performing calculations.

renewable energy. Collecting power from renewable resources that naturally renew within a human timeframe.

space exploration. Discovering and exploring celestial structures using continuously developing space technology.

technology. Application of science for practical outcomes.

telecommuting. Working from home, making use of the Internet, email, and telephone.

telemedicine. Utilizing telecommunications technology for remote diagnosis and treatment of patients.

Universal Basic Income (UBI). A framework for granting every citizen in a country or geographic region a specified amount of money, regardless of their resources, employment situation, or income.

Unmanned Aerial Vehicles (UAVs). An aircraft without a human pilot on board.

virtual reality (VR). A simulated experience that can be similar to or entirely different from the real world.

wearable technology. Electronic gadgets specifically designed to be worn on the body, providing advanced features and capabilities.

xenotransplantation. The process of transplanting cells, tissues, or organs from one species to another.

yield (in agriculture). The production amount per unit of land.

zoonotic diseases. Diseases that spread from animals to humans.

www.ingramcontent.com/pod-product-compliance
Lightning Source LLC
Chambersburg PA
CBHW041933260326
41914CB00010B/1282